Ashraf El-Shamy
Samar M. Mouneir

The Recent Techniques for Corrosion Monitoring in Petroleum Industry

Ashraf El-Shamy
Samar M. Mouneir

The Recent Techniques for Corrosion Monitoring in Petroleum Industry

Destructive and Non-Destructive Testing

Noor Publishing

Imprint
Any brand names and product names mentioned in this book are subject to trademark, brand or patent protection and are trademarks or registered trademarks of their respective holders. The use of brand names, product names, common names, trade names, product descriptions etc. even without a particular marking in this work is in no way to be construed to mean that such names may be regarded as unrestricted in respect of trademark and brand protection legislation and could thus be used by anyone.

Cover image: www.ingimage.com

Publisher:
Noor Publishing
is a trademark of
International Book Market Service Ltd., member of OmniScriptum Publishing Group
17 Meldrum Street, Beau Bassin 71504, Mauritius

Printed at: see last page
ISBN: 978-620-0-07627-4

The Recent Techniques for Corrosion Monitoring in Petroleum Industry

By:

Ashraf El-Shamy

Professor of Physical Chemistry, National Research Centre,

Cairo, Egypt

Samar M. Mouneir

Prof of Pharmacology, Faculty of Veterinary Medicine, Cairo

University

Contents

Methods of NDT

Visual

Tap Testing

X-ray

Acoustic Emission

Ultrasonic

Flux Leakage

Microwave

Magnetic Measurements

Laser Interferometry

Thermography

Magnetic Particle

Acoustic Microscopy

Liquid Penetrant

Replication

Eddy Current

The Recent Techniques for Corrosion Monitoring in Petroleum Industry

A. M. El-Shamy

National Research Centre, Physical Chemistry Department, Electrochemistry and Corrosion

Laboratory, El-Bouhus Street, 12622, Dokki, Cairo, Egypt

E-mail: elshamy10@yahoo.com

Samar M. Mouneir

Department of Pharmacology, Faculty of Veterinary Medicine, Cairo University, Giza, Egypt,

12211

E-mail: samar_mouneir@yahoo.com; samar_mouneir@cu.edu.eg

Abstract

The corrosion monitoring techniques are basically used to monitor or detect the effects of the occurred corrosion. These categories of monitoring techniques basically classified into seven techniques CEION, Electrical Resistance Monitoring, Electrochemical Methods, Hydrogen Monitoring, Weight Loss Coupons, Non-Destructive Testing (NDT) Techniques, Analytical Techniques. The corrosion monitoring techniques are essintial for gathering the corrosion data which facilitate and improved the processes of controlling the damages resulting from the corrosion faliure. There are many and different techniques used for monitoring the corrosion bahavior some of them are destructive and some are non- destructive. The corrosion processes in industry are classified into internal and external corrosion, in case of internal corrosion the corrosion inhibitors are used to control the corrosion process in many industries. The internal corrosion could be monitored by many techniques such as; electrochemical and weight loss methods and these two methods are considered as the back bone of corrosion monitoring for internal corrosion. It is well known that electrochemical methods constitute open circuit potential, potentiodynamic polarization, potentiostatic polarization, cyclic polarization, cyclic volatametry and electrochemical impedance spectroscopy.these techniques provide the corrosion rate and corrosion behavior but it can not introduce the possibility of corrosion and corroion control mechanism, so it needs more techniques to help knowing the mechanism of corrosion inhibitors as scanning electron microscopeand electron diffraction X-ray (EDS). In the field or which is called practical corrosion monitoring, there are many or different programs or techniques could be used for corrosion monitoring depending on the situation, condition and the selected materials. The most used and preferred techniques are weight loss and CEION or ER monitoring because of these methods provides more useful and approximately actual information on the corrosion behavior and easily monitored day by day without any problems which faced the other techniques.

Key words:

Corrosion Monitoring Techniqes; Destructive Methods; *Ceion; Electrical Resistance Monitoring; Electrochemical Methods; Hydrogen Monitoring; Weight Loss Coupons; Non-Destructive Testing (Ndt) Techniques; Analytical Techniques*

1. Introduction

1.1. What is meant by corrosion monitoring?

The definition of corrosion monitoring is attributed with the tools which assigned to the measurement or observation of corrosion phenomenon. The tools of corrosion monitoring have been ranging and classified depending on the nature, conditions and the type of corrosion. Finally, it could be concluded that the corrosion monitoring is the method of measuring or observation the corrosion progress in certain time. The corrosion monitoring techniques are covering the different materials depending on the instrumentation which supported by the different standards and achieve the calculations and data analysis. By measuring the corrosion rate the corrosion behavior could be understand and the corrosion control became easier and the monitoring methods could be improved [1-11]. The inspection is concerned with the conditions which affected on the corrosion process at certain time but the monitoring process is subjected to obtain more information about the corrosion conditions through specific time, so the change in corrosion conditions with times considered as the basic role of monitoring the corrosion process. The corrosion rate is used as function of determining corrosion damage and help the corrosion specialist to understand the safe and useful operation and determines the process duration time [12-19]. The continuous monitoring of the corrosion process could be achieved by using sensors, but it is very important to be very sensitive and enough to give ideal control for the corrosion process. The corrosion control showed clearly the answers of three famous questions the first one is where exactly the corrosion damage is occurred, the second question is when the damage could be existing and finally why the damage is exhibit. The activity of corrosion is proportional to many factors depending essentially on the conditions around the metallic structures [20-26]. The corrosion process is increases by increasing the temperature and the duration time and in the same time the salinity of water is considered the highest effecting factor because of the contents of chloride and/or sulfide ions. The internal corrosion process could be controlled by adding organic compounds to isolate the metallic structures from the environment and this method is monitored by direct techniques. Most of these techniques are used in broad sector of applications and some of them are used in narrow applications depends on the environment of corrosion processes and the field situations. Some of them designed for using in laboratory and the other designed for field monitoring, so there is a specific difference in accuracy between the laboratory and field measurements. The corrosion monitoring techniques could be summarized in three categories as follows:

A- Non-electrochemical techniques

Weight loss method

Gasometrical method

Electrical resistance method

B- Electrochemical and special methods

Linear polarization method

Potential measurements method

Polarization measurements method

Impedance measurements method

C- Non-destructive testing methods

Ultrasonic

Eddy current method

Radiographic method

2. Techniques of corrosion monitoring

2.1. Non-electrochemical techniques

2.1.1. Weight loss method

The monitoring technique which is depending on the weight loss coupon system is considered as the best technique for measuring the corrosion rate of the metallic structures by using coupon or metal specimen and exposing it to an corrosive media and measuring the resultant weight loss for a given duration time. The coupons form depending on the aim of the test and also depending on the corrosive media and the selected metallic structures, so this method is not an instrumental method. The following figure shows the different forms of coupons and also the suitable holders [27].The monitoring of corrosion rates is very important in any environment is due to it is shows the best method for maintenance and/or repair the damaged parts and expecting the costs associated with failure based-corrosion. The test coupons are illustrating the actual corrosion rate with low cost method for measuring the internal corrosivity of the system. The corrosion rate (mpy) information for the used exposed coupons can introduce the life expectancy for the system's materials [28]. The best benefits of this technique are easy to select the coupons materials, size and shape depending on the system's design. The obtained corrosion results can identify the elemental compositions of the used materials and provide this information to the salesperson to select the proper type of coupons materials. As we mentioned before the weight loss process is achieved by using corrosion coupons, this method is very simple and considered as the actual and accurate method for estimating the corrosion rate by calculation of weight loss in the used coupons. The weight loss is calculated at certain time or at time intervals, the coupons are weight before and after treatment were calculated. The corrosion products on the corrosion coupons have to remove and then cleaned and rinsing before reweighting, then the corrosion rate is calculated by using proper equation. One of the benefits of calculating the corrosion rate is overcoming the failures in the infrastructures and plant machines which save the repair cost and prevent the damage of environment and so improve the safety for human. The second benefits of corrosion monitoring are expecting the lifetime or durability of the tested materials. The third benefits of the corrosion monitoring is providing the cost of corrosion damage and where the expecting damage so it provide the early alarm for failures [29].To calculate the corrosion rate (CR) or a metal loss (ML), the test coupons are weighed before testing and after experiment the coupons then cleaned from all corrosion products and then reweighed and the weight loss is could be converted to corrosion rate as follows:

$$\text{Corrosion Rate (CR)} = \frac{\text{Weight loss (g) * K}}{\text{Alloy Density (g/cm}^3\text{) * Exposed Area (A) * Exposure Time (hr)}}$$

$$\text{Metal Loss (ML)} = \frac{\text{Weight loss (g) * K}}{\text{Alloy Density (g/cm}^3\text{) * Exposed Area (A)}}$$

2.1.1.1. Coupon position and orientation

The surface preparation of working electrode coupons affected by many factors could not be controlled e.g. the defects which may be occurred in the microstructure of the metallic coupon. This defect leads to reduce the weight loss accuracy; therefore, we use more than one electrode may be duplicate or multi replicate coupon samples for good accuracy. Another very important factor is the coupon orientation, since it must be consistent to avoid the disturbance in obtained data [30]. The parallel orientation is preferable especially in continuous flow experiment to reflect the true conditions. The best way to optimize the orientation of the used coupons is using coupon holders with automatic alignment during the flow conditions. The position of working electrode coupons is considered as critical factor because the wrong position may cause increase in the corrosion rate and to overcome this problem, we can use ladder-strip coupon holder as mentioned in the following figure.

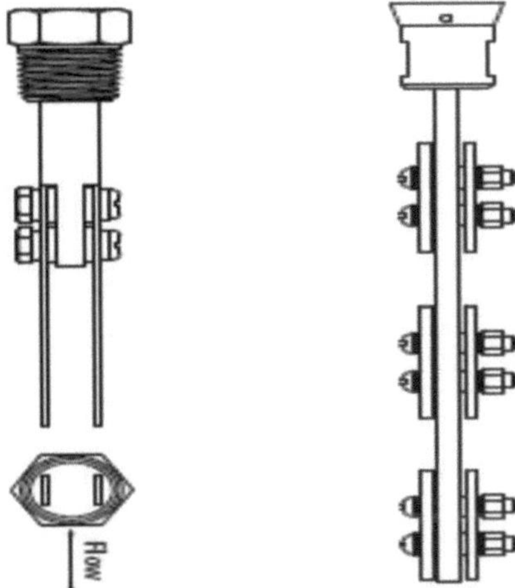

Preferred flow direction   **Ladder strip coupon holder**

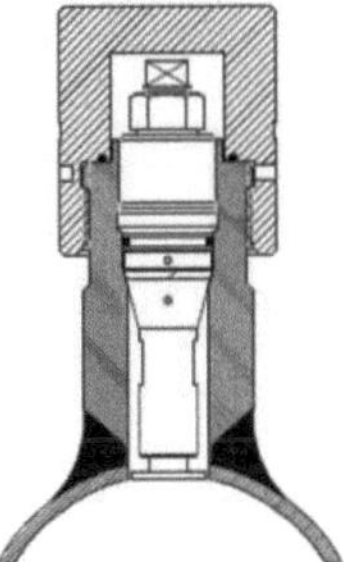

Flush disc coupon holder in high pressure access fitting

The flush-disc coupons are preferable in this situation since the coupon orientation and positioning become in relation to flow. The design for coupon holders becomes very important to avoid the orientation and positioning of the coupons as coupon holder. The size and the type of this holder are generally proved the limiting factor for the configuration of the used coupon. To avoid the mistakes in calculation of the corrosion rate the holders must be constructed from 316L stainless steel which have good corrosion resistance especially in severe conditions e.g. rating of the pressure is up to 3000 psi, and the rating of the temperature is up to 450 °F / 232 °C. The by-pass loop is advised to use the following design holder.

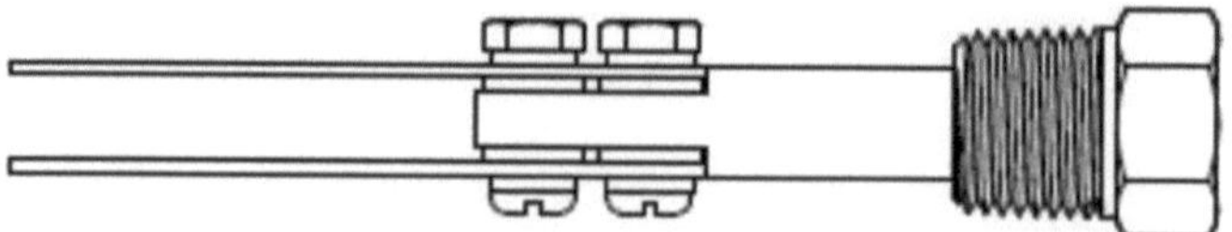

Fixed (pipe plug) coupon holder

There is retractable type for coupon holder is highly recommended to use it in petrochemical industry and/or petrol refining, since this holder design is easy to set up and removal, through a ball-valve, without shut down for the system. These holders can be used in high pressure up to 1500 psi and high temperature up to 500 °F (260 °C) and it is mentioned in the following figure.

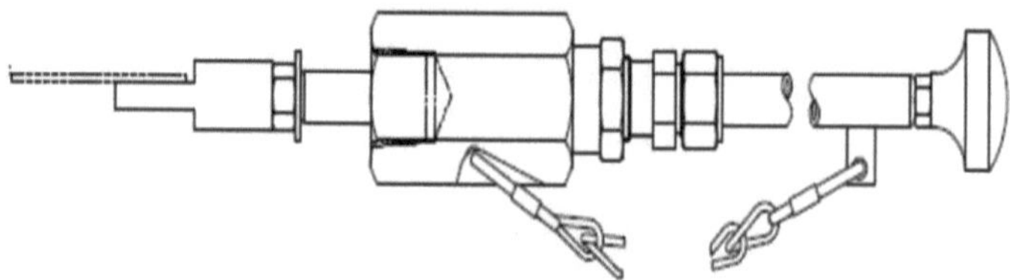

Retractable coupon holder

There are many designs for coupons depending on the conditions and the situations see the following kinds and shapes.

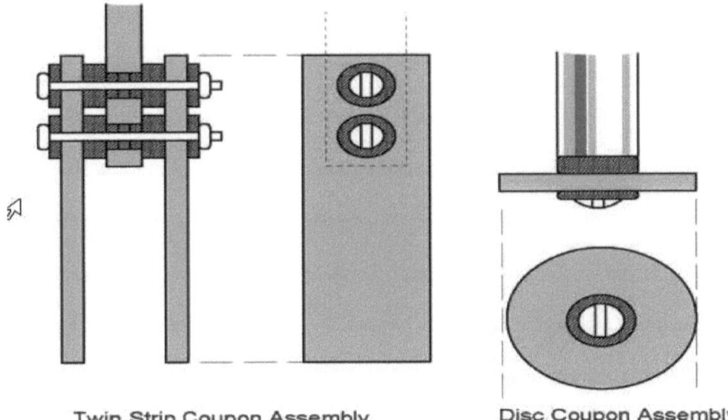

There are special purposes coupons depending on the kind of experiments and in the same time its aim depending on the standard methods as piping systems and stressed coupons. The strength, ductility, and hardness of used materials and also the evaluation of physical characteristics are could be illustrated by the mechanical test specimens (MTS), these data are very important to help the scientist or engineer to collect the important information to select the suitable materials for the working conditions in the field. The following figures illustrate a wide variety of sizes and materials for selection the requirements for test specimens to evaluate the fracture mechanics of the metal samples according to ASTM standard.

2.1.2. Gasometrical method

2.1.2.1. Hydrogen flux monitoring

The hydrogen monitoring technique is used for detecting the provided hydrogen which is considered as indication for corrosion damage. This technique is highly recommended for monitoring the corrosion in plant, which is exposed to acidic environment, since the expected generation of hydrogen gas is exists. It is very good monitoring technique for wet sour gas hydrogen sulfide (H_2S) condition or acidizing in drilling process because of under these conditions the evolved hydrogen gas is absorbed into the metallic structures. It is well known that the produced hydrogen gas cause embitterment, stress corrosion cracking and blistering. The diffusion of hydrogen gas could be monitored through the metallic structures by using the hydrogen monitoring probes. The probes are attached to the exterior of the plant or it inserted to the metallic structures to monitor or measure the presence of hydrogen gas. It already knows that the hydrogen gas or atomic hydrogen is produced as results of corrosion process as a corrosion product in acidic or neutral medium. This atomic hydrogen is used for monitoring the corrosion purposes so hydrogen monitoring sensors could be used by attaching it to the outside walls of tanks or pipelines. The hydrogen sulfide already exists in the petroleum industry, and presence of this gas promotes the hydrogen gas uptake into the steel structure. There are three essential factors which affected on the probes of hydrogen monitoring.

1- **Pressure**

In the controlled chamber the pressure is increases with time, as the hydrogen gas passes through the material into the probe chamber.

2- **Resulting current**

The applied potential cause oxidation of hydrogen and then the current is resulted.

3- **Current flow**

The current flow in an external circuit is depending on the principle of the fuel cell, since the hydrogen gas entering the fuel cell and causes the current flow.

The pressure is one of the important factors so the high pressure hydrogen probe is the best method for measuring the corrosion behavior in high pressures up to 1500 psi, so the probe is consists of three assemblies the first is the gage assembly, the second is the insertion of rod sensing technique, and the third is the packing gland.

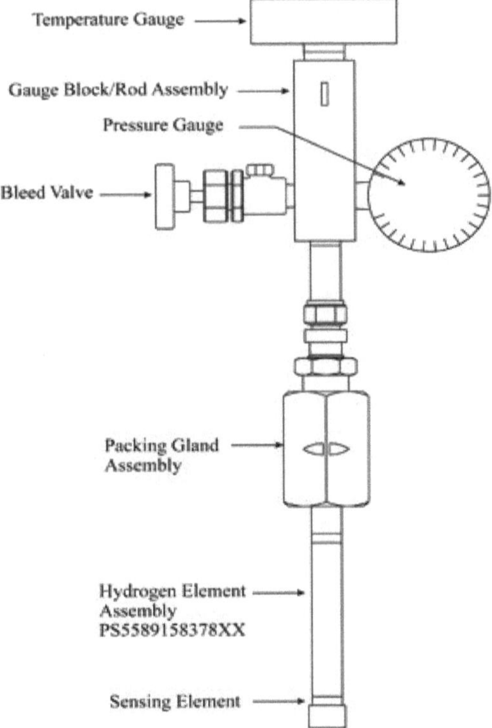

Hydrogen probe retractable with packing gland

2.1.2.2. Hydrogen penetration monitoring system

The technique of Hydrogen Penetration Monitoring System is considered as the best method for measuring the migration rate of hydrogen gas directly through the tanks or the wall of the pipelines. The word through means that the penetrated hydrogen gas inside the walls of tanks or pipelines, so by this technique we can measure the penetration current which results from moving the hydrogen gas through the steel vessels. The penetrated hydrogen could be measured by using the hydrogen patch probe by mounting it to the outside of the pipelines wall by mechanical straps. In this technique there is no need for welding or taping and this one of its benefits because the probe can be simply moved and installed to other location in short time. The following figure shows the Hydrogen Penetration Monitoring System [34-38].

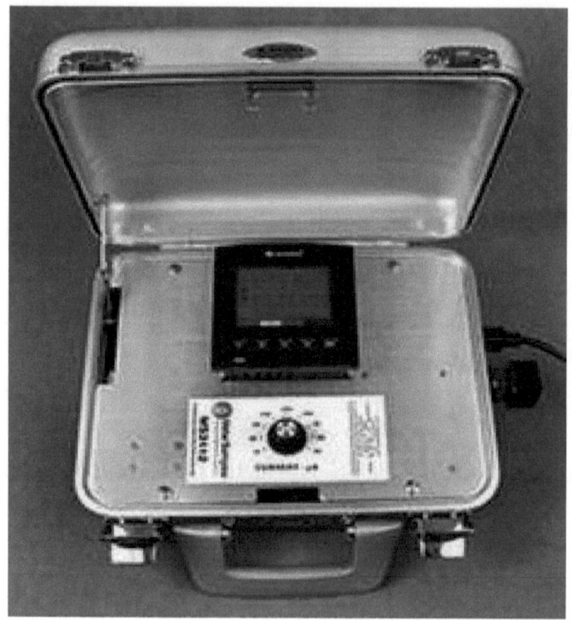

Hydrogen penetration monitoring system

2.1.2.3. Electrical resistance method

The electrical resistance monitoring techniques (ER)are considered as the most applicable and widely use spread techniques because of it is depending upon the determination of resistance which could be changed as a function of corrosion of metallic structures due to the environmental effect. The role of this device is attributed with the proportional relationship between the corrosion rate and the change of resistance in a certain time of exposure. The obtained data from this technique could be adapted with any corrosive media because the probes of electrical resistance are simple to use, and the gained data are well interpreted and easy to prove in applications [39]. The corrosion processes could be monitored continuously for one or more probes so we can study the factors which affect and related to the corrosion processes. This technique has additional benefits because it can be used in monitoring the corrosion rate in different media as liquid, gas in stagnant and continuous flow systems, so it could be considered one of the primaries online monitoring techniques. The electrochemical resistance technique is suitable for the following applications:

- Transmission systems

- Production of oil and gas

- Refinery

- Streams of petrochemical process

- Buried pipeline's external surfaces

- Systems of feed water

- Stacks of flue gas

- Architectural structures

2.2.   Electrochemical methods

It is well known that the corrosion process is considered as an electrochemical process, so it is noticed that there are many monitoring techniques depends on the electrochemical methods. Two famous electrochemical techniques are broadly used in monitoring of the corrosion behavior and tell us about the rate of corrosion in different corrosive media. The fist technique is Linear Polarization Resistance (LPR) and the second technique is the galvanic monitoring and it is also known as (Zero Resistance Ammetry) [40].It is well known that the corrosion rate could be calculated by the electrochemical methods but it cannot give them any information about the weight loss or metal loss. The measurements provided by the electrochemical resistance could be used to obtain the information about the metal loss and this considered as the essential difference between the electrochemical method and the techniques of electrochemical resistance.

2.2.1. Linear polarization method

In these electrochemical techniques the working electrode could be subjected to a small rage of voltage approximately from 10 to 30 mV depending on the metal and the obtained currents are measured. The linear polarization resistance technique is used on microscopic scale which existing in the plant as a microscopic corrosion cells by measuring the corrosion current which passing between the half cells of the anode and the cathode. The produced data are depending on the ratio between voltage and current, which give the polarization resistance due to the mathematical relations between voltage, current and resistance.

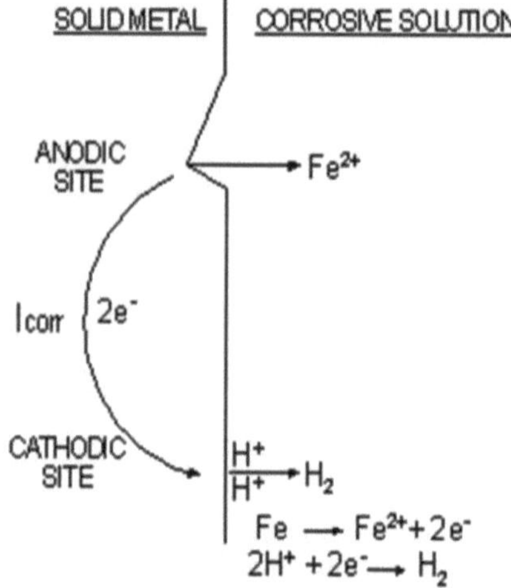

A simple example, of iron dissolving in acidic solution

The increment of the polarization resistance is easily detectable from the decreasing of the corrosion rate, so the monitoring of the linear polarization resistance provides the measure of the rate of corrosion and so it is could be used to optimize the recommended dose of the injected

corrosion inhibitors. From the properties of the linear polarization resistance we can conclude that

this technique is highly restricted to aqueous solutions and the best results depending on the

conductivity of the solution i.e. the good conductive media provide the best results. The accuracy

of the results in higher resistivity solutions is achieved by using three electrode probes working,

counter and reference. The LPR technique is still used till the present especially in aqueous

corrosive environment.

2.2.1.1. Some of the more common applications are:

1- Open and closed cooling water systems

2- Secondary recovery system

3- Distribution systems and potable water treatment

4- Amine sweetening

5- Systems of wastewater treatment

6- Mineral extraction and pickling processes

7- Paper manufacturing

8- Production of hydrocarbon

The LPR technique could be occurred by using two similar electrodes and in this case, it is called

(two electrode system), or the measurements could be made in presence of (three electrode system)

working, reference and counter [41]. The process is depending on immersion of metal or metal

alloys in a conducting solution in presence of sufficient oxidizing power; the metal start to corrode

due to electrochemical reactions (anodic and cathodic electrochemical reaction) the metal surface

is corroded and consumed by anodic reaction by oxidation process. The corrosion rate could be

computed by the measurement of corrosion current results from the flow of electrons from the anodic area. The corrosion rate is calculated by using the following modified equation of Faraday's Law:

$$C = \frac{I_{CORR} \times E}{A \times D} \times 128.67$$

Where:

C = Corrosion rate in "mils per year" (mpy)

E = Equivalent weight of the corroding metal (g)

A = Area of corroding electrode (cm^2)

d = Density of corroding metal (g/cm^3)

The following derived Stern Geary equation give the theoretical relationship for the values of potential and current (E and I) and these values are directly proportional to corrosion current (I_{CORR}), which used as a function of the corrosion rate.

$$\Delta E / \Delta I = \frac{ba\ bc}{2.3\ I_{CORR}\ (ba + bc)}$$

$$I_{CORR} = I/E \times constant$$

The values of (b_a) and (b_c) are constitute the rate constants, the (Tafel) constants, so the equation became:

The experimental calculation became easier by placing a second electrode (auxiliary) in the test solution and connecting it to the working electrode by external power supply. This is illustrated in following Figure:

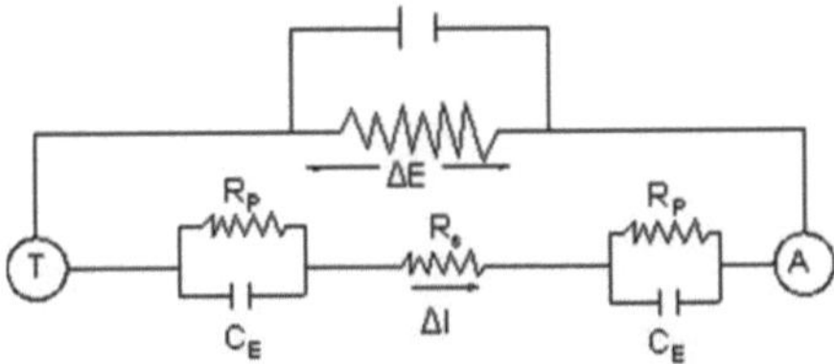

The electrochemical measurements could be used in broad spectrum applications in both the field or in laboratory when it considered that the occurred corrosion is the uniform or general corrosion. Precautions are needed in both linear polarization and (Tafel) extrapolation to achieve the widespread applications of the electrochemical techniques without complications. We can summarize the obstacles in electrochemical techniques as the following:

2.2.1.1.1. Effect of scan rate:

The amount of produced currents is affected by scan rate at any potential value so the scan rate should be selected carefully to avoid misinterpretation of the obtained results. Finally, the scan rate is considered a very important experimental factor so it must be controlled.

2.2.1.1.2. Effect of solution resistance:

To minimize the solution resistance effect, you have to establish the distance between the salt bridge and the working electrode for all experiments and the distance should be limited in all polarization measurements. The effect of solution resistance is clearly noticed in case of solutions with higher resistivity.

2.2.1.1.3. Changing surface conditions:

The surface area of the working electrode should be constant to avoid the change of potential due to the change in the surface area. The corrosion reaction is directly occurred on the electrode surface, so it is very important to take the precautions during the processing of the working electrode to keep the potential stable in reproducible experiments. The localized corrosion could be monitored through the appearance of a hysteresis between the forward and reverse loop of polarization curves and this behavior prove the presence of pitting or crevice corrosion.

2.2.2. Potential measurements method

The corrosion potential measurements are used in a wide application range in industry especially for monitoring the corrosion in reinforcing concrete steel and cathodically protected pipelines [42]. The open circuit potential is used with higher accuracy of anodic and cathodic protection systems, since the corrosion potential changes can give the indication of corrosion behavior through activation and/or passivation. The corrosion potential can also give the indication of the thermodynamic corrosion risk.

2.2.3. Polarization measurements method

The potentiodynamic polarization method is considered as one of the most important corrosion testing and extremely used in laboratory. This method produce very clear and significant information about the process of corrosion mechanisms and in the same time provide the magnitudes of the corrosion rates, which attributed with the relation between the materials and environmental effects [43]. We can conclude that the polarization methods results from the change in the working electrode potential and in the same time monitor the current values, which is obtained as results of time at certain potential. The potentiodynamic polarization method produces data from anodic, cathodic and cyclic polarization which could be illustrated as follows:

2.2.3.1   Anodic polarization:

The expression anodic polarization means that the working electrode shows a significant corrosion rate due to the change of potential takes place on the anode surface in the positive direction.

2.2.3.2.   Cathodic polarization:

The cathodic polarization means that the potential of the working electrode surface change in more negative direction because of electrons addition on the metal surface, which saved the working electrode from corrosion results from the corrosive environment by forming protective layer by electro-deposition process.

2.2.3.3. Cyclic polarization:

This technique is essentially used to detect the localized corrosion by carrying out the polarization in both anodic and cathodic directions. The cyclic polarization is extinguishable to monitor the susceptibility of pitting corrosion since the hysteresis is examined by calculating the area under curve from one cycle by registering the potential starting from the open circuit potential and the passivation potentials.

2.2.3.4. Cyclic voltammetry:

The cyclic voltammetry occurred by passing the potential in a positive direction to a certain value of current or potential and then, the scan is reversed to the more negative values till reached the started potential value.

2.2.3.5. Galvanic Monitoring (Zero Resistance Ammetry)

This technique is depending on the flow current between two metals due to the difference in the electrode potential between these two metals if it externally connected. The obtained current constitutes the dissolution rate of the more negative working electrode to more positive surface [44]. The measurements are related to the oxygen ingress into the protected systems due to the de-aeration process, so it is the best and suitable technique in monitoring the corrosion behavior in flow conditions.

2.2.3.6. Electrochemical potentiodynamic reactivation (EPR):

The EPR technique is suitable for measuring the sensitization degree of stainless steels by sweeping the potentials from passive to active so it called reactivation [45].

2.2.3.7.   Potentiostaic method:

In this technique the working electrode is subjecting to polarization at different values of potential at constant time and the current can stabilize the potential to the next step [46].

2.2.4.   Impedance measurements method

The Electrochemical Impedance Spectroscopy (EIS) is proven to be accurate technique for measuring the rate of corrosion. The polarization resistance in this technique is calculated by the charge transfer resistance, which proportional to the rate of corrosion [47]. The interpretation of the electrochemical impedance spectroscopy could be achieved from knowing the model of interface by applying it on the results from the monitoring the interface. The difference between the electrochemical impedance spectroscopy and the other electrochemical techniques is in case of electrochemical impedance spectroscopy we can use narrow amplitude signals without disturbing the polarization resistance values. Generally, the used voltage is between (5–50 mV) and the frequency range is between 0.001-100000 Hz, these ranges give the chance for the electrochemical impedance spectroscopy to record the real resistance and imaginary capacitance as a response of the EIS system. The interpretation of the electrochemical impedance spectroscopy is also depending on the spectrum shape and the description of the circuit. The circuit is described and designed by fitting the best frequency formed from the EIS spectrum and the fitting is achieved by judging the overlap on the original spectrum. The fitting will help us to obtain the correlated parameters with the condition of coating and also the corrosion behavior of the working electrode.

2.2.5. Electrochemical frequency modulation (EFM)

The Electrochemical Frequency Modulation (EFM) technique for corrosion monitoring is depending upon using the signals of two sinusoidal potential on the working electrode [48]. By applying these potentials, the currents resulting are measured with a period of time, then these obtained data are converted to frequency which help us to measure the signal.

2.2.6. Electrochemical Noise (EN)

The electrochemical noise (EN) monitoring technique for corrosion is the best technique for monitoring corrosion of aircraft and the tower of gas scrubbing. The evaluation of (EN) as a corrosion tool is depending on the fluctuations of potential or current of the corroding metallic specimen. There are many kinds of corrosion could be monitored by using the electrochemical noise technique as; exfoliation, erosion-corrosion and localized corrosion such as stress corrosion cracking(SCC) and pitting which characterized the localized corrosion. The monitoring of these kinds of corrosion is could be applied in either the field environment or in the laboratory and it is proved that the electrochemical noise (EN) monitoring technique is more sensitive than any other electrochemical technique. One of the benefits of electrochemical noise (EN) monitoring technique is easily monitor the passivation breakdown and repassivation processes which results from the film formation or pit propagation processes [49]. The monitoring of corrosion behavior of coatings by Electrochemical Noise Analysis (ENA) is depending on using three electrodes, of which one is working and two are reference. The technique is carried out by measuring the current between the two working electrodes at known period of time. The data of current and potential are used to calculate the corrosion resistance by the following equation:

$$Rn = \frac{StdevE}{StdevI}$$

Where: Stdev E is the standard deviation of the potential

Stdev I is the standard deviation of the current

And the pitting index ratio is calculated by knowing the noise resistance as:

$$PI = \frac{StdevI}{RMSI}$$

Where: RMSI is the root mean square value of the measured current i

The type of corrosion which present on the metallic sample can be determined by knowing the pitting index ratio.

2.2.7. Electrical field Signature method (FSM)

The technique of electrical field corrosion monitoring has the ability to measure the corrosion damage via the metal loss. These techniques can provide the corrosion data over several meters of pipeline surface, so it is suitable or directed for monitoring the corrosion in oil and gas and petroleum production industry. The (FSM) technique is depending upon measurement of the distribution of resulting voltage due to feeding of induced current and this help us to detect the corrosion. The design of electrical field signature corrosion monitoring technique is simply achieved by installation permanent pin around the external surface of the pipeline to monitor the voltage and produce the potential map for the tested structure. The only drawback of this electrical field signature corrosion monitoring technique is the pin spacing since the increasing the pin spacing leads to decrease the resolution especially for the localized corrosion [50].

2.2.8. CEION

The CEION monitoring technique is the best choice device for sub-surface monitoring technology for corrosion damage. It is also highly recommended for monitoring the sub-sea because of its quick response and it is not required maintenance in case of operating and shut down processes. The CEION technique is depending essentially on the measurement of metal loss ratio with very high resolution compared with the electrical resistance depending devices so it is considered as the ideal technique for measuring the corrosion of producing systems results from sand erosion [51]. This monitoring technique has more benefits because of the applicability to set of many sensors, which designed to collect enough data to develop the application processes.

2.3. Non-destructive testing methods

The non-destructive technique includes ultrasonic, radiography, thermo-graph, eddy current measurement.

2.3.1. Ultrasonic

Ultrasonic waves have the ability to travel for a long distance along pre-stressing strands when those strands are suspended in air. The signal is travelling to limit distance for about one or two meters when the strands are embedded in concrete. The grain size of the aggregates is affecting the signal of ultrasonic wavelength which leads to large wave scattering. The analysis of ultrasonic signals for the purpose of assessing the condition of pre-stressing steel is depending on the frequency of steel strand and it is can be calculated by analyzing the input and output signals [52].The frequency response of strand is very sensitive to small amounts of corrosion. The multiple frequency band correlation technique is used for analyzing the receiving signals.

2.3.2. Eddy current method

One of the non-destructive techniques is pulsed eddy current method; it is used for monitoring, inspection and identification of hidden corrosion. This technique is suitable for monitoring the corrosion in layered as aircraft and the thickness of conductive coatings on metal plates. Due to the pulsed excitation this technique able to cover wide range of frequencies quickly and this is considered the major advantage of this technique [53]. The other advantages are concentrated on the cost, since it is inexpensive technique and the manual is very simple to use [54]. The eddy current technique depends on the theory of electromagnetic induction, so it penetrates into the layers of subsurface even these layers are not mechanically bonded. The electromagnetic induction means the current induced by induction, this technique depends on the measure the change of flow current which induced by the induction coil. The difference between this technique and the ultrasonic technique is this technique can monitor the corrosion where mechanical contact between layers is required. This technique is also capable to detect the metal loss in 2-3 layers and have the ability to differentiate between metal loss and metal separation.

2.3.3. Radiographic method

2.3.3.1. Thin layer activation (TLA)

The field of nuclear science help the scientists to create a new monitoring technique for corrosion in this technique a small area of metallic material is exposed to charged particles of a beam with high energy to form a radioactive layer on the surface. In this technique the proton beam is used to produce radioactive isotope inside the metallic steel surface and give the chance to the isotope to decay during the emission of gamma radiation. The monitoring of metallurgical properties of the steel structures are unchanged due to the lower concentration of radiation, so it is very safe technique and should not compared with the normal radiography [55]. The sensors could be used in this technique by detecting the emission of gamma radiation from the surface layer and the rate of material removal could be measured.

2.3.3.2. Visual Interpretation

The uniform corrosion or the kind or degree of localized corrosion can be observed and measured by analyzing the weight loss of visual inspection. The localized corrosion as pitting corrosion is easily noticed by naked eye and could be proved by surface characterization via optical or scanning electron microscope.

2.3.3.3. Analytical Techniques

This technique is depending on the analysis of fluid samples as conductivity, temperature, pH and flow measurements, counts of chloride, oxygen and iron. These data can provide the required information about the corrosion monitoring. The required items for analytical techniques could be monitored by using suitable sensors. Multi-electrode Sensor(CMS) device can do it easily but in many situations, the process status and the product quality are determined by using traditional chemical methods and the localized corrosion monitoring could be measured by electrochemical studies [56].Under normal temperature and pressure or high-temperature and high-pressure conditions, the monitoring of corrosion (localized and/or uniform) is easy achievable in liquids, soils, concrete, and humid gases.

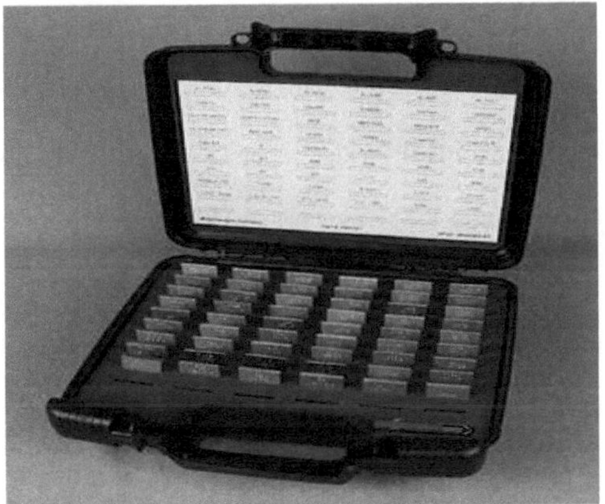

Kits for checking the elemental content

The analytical techniques are depending on using of specific kits can detect the elemental contents of any alloy by examining the intensity of color for any chemical spot in comparison to a standard. The used kit of can be subjected for testing and could be compared with the metal standard to ensure the shelf life. The chemical analysis must be certified for each and recorded on an analysis sheet.

2.3.3.4. Inductive resistance probes

In this technique the change of inductive resistance of a coil is used to measure the changes in located inside the sensor element. The sensors are subjected to improvement to make it not like the electrochemical resistance probes and make the sensors are could be used in a widely range of environmental conditions [57]. The advantage of this technique is the ability to measure the corrosion rate in non-aqueous and low conductive environment and the electrochemical techniques is unsuitable in these conditions.

2.3.3.5. Harmonic analysis for corrosion monitoring

This monitoring technique for corrosion is used in large scale in the field and it is considered one of the related techniques to electrochemical impedance spectroscopy but in this technique the potential is applied to one sensor element in a three element probe and the obtained current is detected. In this technique the harmonic oscillations are detected with the primary frequency and no other technique offers this facility [58]. This technique has a unique feature, since this technique will be developed to create a tool for corrosion monitoring in cathodic protection systems. More benefits of this technique can be applied in a broad range of frequency and all the polarization resistance, Tafel parameters and corrosion rate could be measured.

2.3.3.6. Acoustic emission (AE)

The sound waves are generally produced from the mechanical stresses generated during the changes of temperature or pressure. The growth of microscopic defects of stress corrosion cracks emits acoustic sound waves; these waves could be measured by acoustic emission monitoring techniques, which indicated to the corrosion behavior [59]. The sensors could be used in this technique by selecting the suitable position in the metallic structures.

2.3.3.7. Multi electrode sensor analyzers

The metal corrodes less or does not corrode depending on the electrons transfer from anodic are to cathodic area, this behavior is already existing in presence of dissimilar metals and this leads to localized corrosion or non-uniform corrosion [60]. The used sensor in this technique is depending on multi-electrodes made from identical materials to the engineering components. Some of these electrodes compatible to the anodic area and others are compatible to the cathodic area depending on the material properties of the corroding metal. Finally, the resulting electrical currents are could be measured and the kind of corrosion especially the localized corrosion rates are determined.

Multi electrode Sensor Analyzers

2.3.3.8.   High resolution ER data logger

The corrosion meter device is perfect in measuring and storing data from the most types of probes, which detected the electrical resistance and it, is considered as indication for the corrosion behavior and also, we can measure the corrosion rate. The measurement of resistances is changed over different periods of time which can be indicated to the occurred metal loss. There is a proportional relationship between the changes of electrical resistance and the corrosion rate.

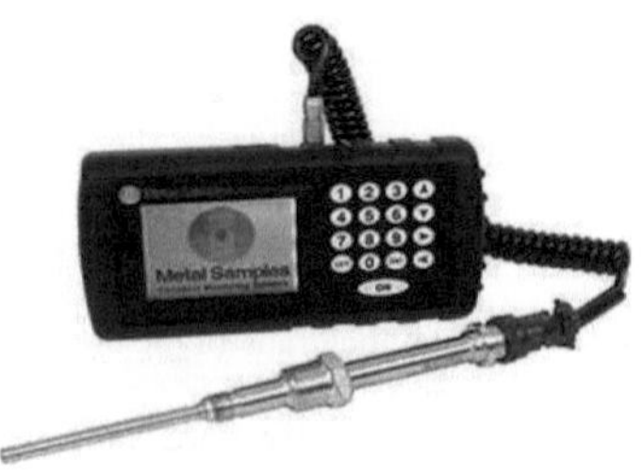

High resolution ER data logger

This technique can detect the corrosion rate in conductive and nonconductive environments, so it is finding a wide variety of applications. Different industries can be subjected to monitor the corrosion by this device such as; petroleum companies, chemical industries, water and wastewater plants, soil corrosivity, or even atmospheric corrosion. The sensitivity and accuracy of this technique is very high, so it can detect the metal loss even in a small increment and the response is faster than the traditional electrochemical resistance techniques [61-63]. The obtained data can then be stored to external mass memory and for more explanations, all stored readings are automatically registered with time, so it is considered as corrosion archive for the plant or factory. The stored corrosion data could be managed by using corrosion software. The analysis of corrosion data can be imported into any standard data analysis because and data can be opened and charted using excel sheet.

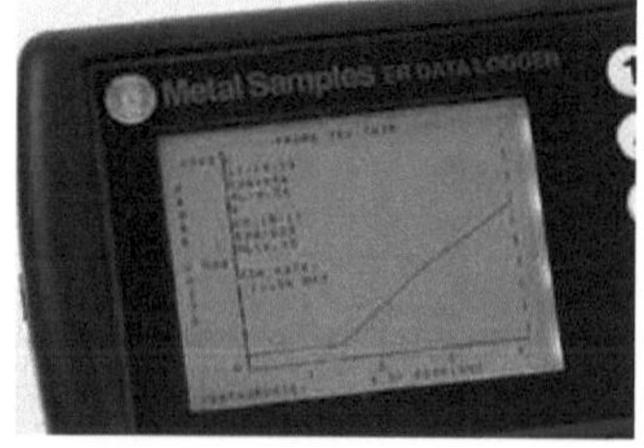

On-screen charting Transfer data directly to USB Flash drive

References

1. M. F. Montemor, A. M. P. Simoes, M. G. S. Ferreira. "Chloride-induced corrosion on reinforcing steel: from the fundamentals to the monitoring techniques". Cement & Concrete Composites 25 (2003) 491–502.

2. Y. Sato_, T. Atsumi, T. Shoji. "Continuous monitoring of back wall stress corrosion cracking growth in sensitized type 304 stainless steel weldment by means of potential drop techniques". International Journal of Pressure Vessels and Piping 84 (2007) 274–283.

3. J. Castrellon-Uribe, C. Cuevas-Arteaga, A. Trujillo-Estrada. "Corrosion monitoring of stainless steel 304L in lithium bromide aqueous solution using transmittance optical detection technique". Optics and Lasers in Engineering 46 (2008) 469–476.

4. I. Martínez, C. Andrade. "Examples of reinforcement corrosion monitoring by embedded sensors in concrete structures". Cement & Concrete Composites 31 (2009) 545–554.

5. Wenyi Liu, Baoping Tang, Yonghua Jiang. "Status and problems of wind turbine structural health monitoring techniques in China". Renewable Energy 35 (2010) 1414–1418.

6. Guofu Qiao, Guodong Sun, YiHong, Yuelan Qiu, Jinping Ou. "Remote corrosion monitoring of the RC structures using the electrochemical wireless energy-harvesting sensors and networks". NDT&E International 44 (2011) 583–588.

7. Mohammed A. Amin *, Gaber A. M. Mersal, Q. Mohsen. "Monitoring corrosion and corrosion control of low alloy ASTM A213 grade T22 boiler steel in HCl solutions". Arabian Journal of Chemistry (2011) 4, 223–229.

8. Junqi Gao, Jin Wu, Jun Li, Xinming Zhao. "Monitoring of corrosion in reinforced concrete structure using Bragg grating sensing". NDT&E International 44 (2011) 202–205.

9. Guofu Qiao, Yi Hong, Guangping Song, Hui Li, Jinping Ou. "Electrochemical characterization of the solid-state reference electrode based on NiFe2O4 film for the corrosion monitoring of RC structures". Sensors and Actuators B 168 (2012) 172– 177.

10. Xingpeng Guo, Zhehua Dong, Tan Gu, Xi Yuan, Xiankang Zhong, Zunyi Yang, Dandan Zhang. "Research on the intelligent safety monitoring system of pipeline corrosion in acidic oil and gas fields II". Procedia Engineering 27 (2012) 1664 – 1670.

11. Pascal Beese, Hendrik Venzlaff, Jayendran Srinivasan, Julia Garrelfs, Martin Stratmanna, Karl J.J. Mayrhofer. "Monitoring of anaerobic microbially influenced corrosion via electrochemical frequency modulation". Electrochimica Acta 105 (2013) 239– 247.

12. Juliusz Orlikowski, Kazimierz Darowicki, Stanisław Mikołajski. "Multi-sensor monitoring of the corrosion rate and the assessment of the efficiency of a corrosion inhibitor in utility water installations". Sensors and Actuators B 181 (2013) 22– 28.

13. A. Brenna, F. Bolzoni, S. Beretta, M. Ormellese. "Long-term chloride-induced corrosion monitoring of reinforced concrete coated with commercial polymer-modified mortar and polymeric coatings". Construction and Building Materials 48 (2013) 734–744.

14. Ch. Thee, Long Hao, Junhua Dong, Xin Mu, Xin Wei, Xiaofang Li, Wei Ke. "Atmospheric corrosion monitoring of a weathering steel under an electrolyte film in cyclic wet–dry condition". Corrosion Science 78 (2014) 130–137.

15. Daisuke Mizuno, Sachiko Suzuki, Sakae Fujita, Nobuyoshi Hara. "Corrosion monitoring and materials selection for automotive environments by using Atmospheric Corrosion Monitor (ACM) sensor". Corrosion Science 83 (2014) 217–225.

16. Pierre-Adrien Itty, Marijana Serdar, Cagla Meral, Dula Parkinson, Alastair A. MacDowell, Dubravka Bjegovic´, Paulo J.M. Monteiro. "In situ 3D monitoring of corrosion on carbon steel and ferritic stainless steel embedded in cement paste". Corrosion Science 83 (2014) 409–418.

17. Hien Nguyen, Thillainathan Venugopala, Shuying Chen, Tong Suna, Kenneth T.V. Grattan, Susan E. Taylor, P.A. Muhammed Basheer, Adrian E. Long. "Fluorescence based fiber optic pH sensor for the pH 10–13 range suitable for corrosion monitoring in concrete structures". Sensors and Actuators B 191 (2014) 498– 507.

18. S. P. Karthick, S. Muralidharan, V. Saraswathy, K. Thangavel. "Long-term relative performance of embedded sensor and surface mounted electrode for corrosion monitoring of steel in concrete structures". Sensors and Actuators B 192 (2014) 303– 309.

19. N. A. Hoog, M. J. J. Mayer, H. Miedema, R. M. Wagterveld, M. Saakes, J. Tuinstra, W. Olthuis, A. vanden Berg. "Stub resonators for online monitoring early stages of corrosion". Sensors and Actuators B 202 (2014) 1117–1136.

20. Shilpa Patil, Bilavari Karkare, Shweta Goyal. "Acoustic emission vis-à-vis electrochemical techniques for corrosion monitoring of reinforced concrete element". Construction and Building Materials 68 (2014) 326–332.

21. Milan Kouril, Tomas Prosek, Bert Scheffel, Yves Degres. "Corrosion monitoring in archives by the electrical resistance technique". Journal of Cultural Heritage 15 (2014) 99–103.

22. J. Izquierdo, B.M. Fernández-Pérez, J. J. Santana, S. González, R. M. Souto. "In situ monitoring of the electrochemical reactivity of aluminum alloy AA6060 using the scanning vibrating electrode technique". Journal of Electroanalytical Chemistry 732 (2014) 74–79.

23. William M. Cox. "A Strategic Approach to Corrosion Monitoring and Corrosion Management". Procedia Engineering 86 (2014) 567 – 575.

24. V. Maruthapandian* and V. Saraswathy. "Solid Nano Ferrite Embeddable Reference Electrode for Corrosion Monitoring in Reinforced Concrete Structures". Procedia Engineering 86 (2014) 623 – 630.

25. Jaka Kovac, Carole Alaux, T. James Marrow, Edvard Govekar, Andraz Legat. "Correlations of electrochemical noise, acoustic emission and complementary monitoring techniques during intergranular stress-corrosion cracking of austenitic stainless steel". Corrosion Science 52 (2010) 2015–2025.

26. Flávio Felix Feliciano, Fabiana Rodrigues Leta, Fernando Benedicto Mainier. "Texture digital analysis for corrosion monitoring". Corrosion Science 93 (2015) 138–147.

27. NaingNaing Aung, Edward Crowe, XingboLiu. "Development of self-powered wireless high temperature electrochemical sensor for in situ corrosion monitoring of coal-fired power plant". ISA Transactions 55(2015)188–194.

28. Heming Wei, Xuefeng Zhao, Dongsheng Li, Pinglei Zhang, Changsen Sun. "Corrosion monitoring of rock bolt by using a low coherent fiber-optic interferometry". Optics & Laser Technology 67 (2015)137–142.

29. Ashutosh Sharma, Shruti Sharma, Sandeep Sharma, Abhijit Mukherjee. "Ultrasonic guided waves for monitoring corrosion of FRP wrapped concrete structures". Construction and Building Materials 96 (2015) 690–702.

30. Hou Bo, He Yuting, Cui Ronghong, Gao Chao, Zhang Teng. "Crack monitoring method based on Cu coating sensor and electrical potential technique for metal structure". Chinese Journal of Aeronautics, (2015), 28(3): 932–938.

31. Guofu Qiao, Yi Hong, Jinping Ou. "Corrosion monitoring of the RC structures in time domain: Part I. Response analysis of the electrochemical transfer function based on complex function approximation". Measurement 67 (2015) 78–83.

32. Guofu Qiao, Yi Hong, Jinping Ou, Xinchun Guan. "Corrosion monitoring of the RC structures in time domain: Part II. Recognition algorithm based on fractional derivative theory". Measurement 67 (2015) 84–91.

33. Huafu Pei, Zongjin Li, Jinrui Zhang, Qian Wang. "Performance investigations of reinforced magnesium phosphate concrete beams under accelerated corrosion conditions by multi-techniques". Construction and Building Materials 93 (2015) 989–994.

34. A. Brenna, L. Lazzari, M. Ormellese. "Monitoring chloride-induced corrosion of carbon steel tendons in concrete using a multi-electrode system". Construction and Building Materials 96 (2015) 434–441.

35. D. R. Vazquez, Y. A. Villagrán Zaccardi, C. J. Zega, M. E. Sosa, G. S. Duffó. "Implementation of Different Techniques for Monitoring the Corrosion of Rebars Embedded in Concretes Made with Ordinary and Pozzolanic Cements". Procedia Materials Science 8 (2015) 73 – 81.

36. Kaige Wu, Woo-Sang Jung, Jai-Won Byeon. "In-situ monitoring of pitting corrosion on vertically positioned 304stainless steel by analyzing acoustic-emission energy parameter". Corrosion Science 105 (2016) 8–16.

37. A.G. Marques, M.G. Taryba, A.S. Panão, S.V. Lamaka1, A.M. Simões. "Application of scanning electrode techniques for the evaluation of iron–zinc corrosion in nearly neutral chloride solutions". Corrosion Science 104 (2016) 123–131.

38. Hong-Qi Yang, Qi Zhang, San-Shan Tu, You Wang, Yi-Min Li, Yi Huang. "Effects of inhomogeneous elastic stress on corrosion behavior ofQ235 steel in 3.5 % NaCl solution using a novel multi-channel electrode technique". Corrosion Science 110 (2016) 1–14.

39. Fabienne Delaunois, Alexis Tshimombo, Victor Stanciu, VéroniqueVitry. "Monitoring of chloride stress corrosion cracking of austenitic stainless steel: identification of the phases of the corrosion process and use of a modified accelerated test". Corrosion Science 110 (2016) 273–283.

40. B. Elsener, M. Alter, T. Lombardo, M. Ledergerber, M. Wörle, F. Cocco, M. Fantauzzi, S. Palomba, A. Rossi. "A non-destructive in-situ approach to monitor corrosion inside historical brass wind instruments", Microchemical Journal 124 (2016) 757–764.

41. Peter B. Nagy, Francesco Simonetti, GeirInstanes. "Corrosion and erosion monitoring in plates and pipes using constant group velocity Lamb wave inspection". Ultrasonics 54 (2014) 1832–1841.

42. T. Balusamy, T. Nishimura. "In-Situ Monitoring of Local Corrosion Process of Scratched Epoxy Coated Carbon Steel in Simulated Pore Solution Containing Varying percentage of Chloride ions by Localized Electrochemical Impedance Spectroscopy". Electrochimica Acta 199 (2016) 305–313.

43. M. Nie, S. Neodo, J.A. Wharton, A. Cranny, N.R. Harris, R.J.K. Wood, K.R. Stokes. "Electrochemical detection of cupric ions with boron-doped diamond electrode for marine corrosion monitoring". Electrochimica Acta 202 (2016) 345–356.

44. Jacek Ryl, Joanna Wysocka, Pawel Slepski, Kazimierz Darowicki. "Instantaneous impedance monitoring of synergistic effect between cavitation erosion and corrosion processes". Electrochimica Acta 203 (2016) 388–395.

45. Hongping Zhu, Hui Luo, Demi Ai, Chao Wang. "Mechanical impedance-based technique for steel structural corrosion damage detection". Measurement 88 (2016) 353–359.

46. C.H. Tana, Y.G. Shee, B.K. Yap, F.R. Mahamd Adikan. "Fiber Bragg grating based sensing system: Early corrosion detection for structural health monitoring". Sensors and Actuators A 246 (2016) 123–128.

47. L. Yang, T. T. Yang, Y.C. Zhou, Y. G. Wei, R. T. Wu, N. G. Wang. "Acoustic emission monitoring and damage mode discrimination of APS thermal barrier coatings under high temperature CMAS corrosion". Surface & Coatings Technology 304 (2016) 272–282.

48. FangjiGan, Guiyun Tian, Zhengjun Wan, Junbi Liao, Wenqiang Li. "Investigation of pitting corrosion monitoring using field signature method". Measurement 82 (2016) 46–54.

49. Bhuiyan M.S.H., Choudhury I.A., Dahari M., Nukman Y., Dawal S.Z. "Application of acoustic emission sensor to investigate the frequency of tool wear and plastic deformation in tool condition monitoring". Measurement 92 (2016) 208–217.

50. M. Criado, I. Sobrados, J.M. Bastidas, J. Sanz. "Corrosion behavior of coated steel rebars in carbonated and chloride-contaminated alkali-activated fly ash mortar". Progress in Organic Coatings 99 (2016) 11–22.

51. Khalil Al Handawi, Nader Vahdati, Paul Rostron, Lydia Lawand, Oleg Shiryayev. "Strain based FBG sensor for real-time corrosion rate monitoring in pre-stressed structures". Sensors and Actuators B 236 (2016) 276–285.

52. V. Maruthapandian, S. Muralidharan, V. Saraswathy. "V. Maruthapandian, S. Muralidharan, V. Saraswathy". Construction and Building Materials 107 (2016) 28–37.

53. D. V. Ribeiro, J. C. C. Abrantes. "Application of electrochemical impedance spectroscopy (EIS) to monitor the corrosion of reinforced concrete: A new approach'. Construction and Building Materials 111 (2016) 98–104.

54. Arash Behnia, Navid Ranjbar, Hwa Kian Chai, Mahyar Masaeli. "Failure prediction and reliability analysis of ferrocement composite structures by incorporating machine learning into acoustic emission monitoring technique". Construction and Building Materials 122 (2016) 823–832.

55. Rosie A. Grayburn, Mark G. Dowsett, Pieter-Jan Sabbe, Didier Wermeille, Jorge Alves Anjos, Victoria Flexer, Michel De Keersmaecker, Dirk Wildermeersch, Annemie Adriaens. "SR-XRD in situ monitoring of copper-IUD corrosion in simulated uterine fluid using a portable spectro-electrochemical cell". Bio-electrochemistry 110 (2016) 41–45.

56. Agata Jazdzewska, Kazimierz Darowicki, Juliusz Orlikowski, Stefan Krakowiak, Krzysztof Zakowski, MaciejGruszka, Jacek Banas. "Critical analysis of laboratory measurements and monitoring system of water-pipe network corrosion-case study". Case Studies in Construction Materials 4 (2016) 102–107.

57. Luigi Calabrese, Massimiliano Galeano, Edoardo Proverbio, Domenico Di Pietro, Filippo Cappuccini, Angelo Donato. "Monitoring of 13% Cr martensitic stainless-steel corrosion in chloride solution in presence of thiosulfate by acoustic emission technique". Corrosion Science 111 (2016) 151–161.

58. Y. Hou, C. Aldrich, K. Lepkova, L.M. Suarez, B. Kinsella. "Monitoring of carbon steel corrosion by use of electrochemical noise and recurrence quantification analysis". Corrosion Science 112 (2016) 63–72.

59. Z. Brytan, J. Niagaj, Ł. Reiman. "Corrosion studies using potentiodynamic and EIS electrochemical techniques of welded lean duplex stainless steel UNS S82441". Applied Surface Science 388 (2016) 160–168.

60. S. Grassini, S. Corbellini, M. Parvis, E. Angelini, F. Zucchi. "A simple Arduino-based EIS system for in situ corrosion monitoring of metallic works of art". Measurement 114(2018) 508–514.

61. Wade Zaluski, George El-Kaseeh, Si-Yong Lee, Mark Piercey, Andrew Duguid. "Monitoring technology ranking methodology for CO2-EOR sites usingthe Weyburn-Midale Field as a case study". International Journal of Greenhouse Gas Control 54(2) (2016) 466–478.

62. K.M Zohdy, A.M El-Shamy, A. Kalmouch, EAM Gad. The corrosion inhibition of (2Z, 2′ Z)-4, 4′-(1, 2-phenylene bis (azanediyl)) bis (4-oxobut-2-enoic acid) for carbon steel in acidic media using DFT.Egyptian Journal of Petroleum 28 (4) (2019), 355-35.

63. M. Shehata, S. El-Shafey, N. A. Ammar, A. M. El-Shamy. Reduction of Cu+ 2 and Ni+ 2 Ions from Wastewater Using Mesoporous Adsorbent: Effect of Treated Wastewater on Corrosion Behavior of Steel Pipelines. Egyptian Journal of Chemistry 62 (9) (2019), 1587-1602

More
Books!

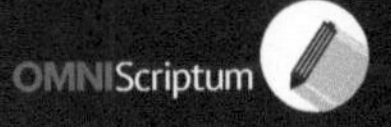

OMNIScriptum

Printed by Books on Demand GmbH, Norderstedt / Germany